IMAGES
of America

SYRACUSE TELEVISION

For more than 27 years, preschoolers across Central New York tuned into *The Magic Toy Shop* on WHEN-TV. The program was created with the help of parents, educators, and other experts to provide a fun and positive experience. Seen here from left to right are Socrates Sampson as Eddie Flum Num, Jean Daugherty as the Play Lady, and Marylin Hubbard Herr as Merrily. (Courtesy Onondaga Historical Association.)

ON THE COVER: Central New York television legend "Baron Daemon" swoops in on a big bunch of "Bloody Buddies" in the early 1960s. Mike Price launched his "clown in a vampire's cape" around Halloween 1962, hosting late movies on "Colorful Channel 9" (then WNYS-TV). Within weeks, demand from viewers forced the station to consider a new time slot, with parents complaining that their children would not go to bed at night until they saw the Baron's funny sketches. "The Baron and His Buddies" was quickly slotted as an afternoon kids' show and lasted until 1967. (Courtesy of Mike Price.)

IMAGES
of America

SYRACUSE TELEVISION

Christie Casciano Burns, Tim Fox,
and Lou Gulino

ARCADIA
PUBLISHING

Published by Arcadia Publishing
Charleston, South Carolina

Printed in the United States of America

Library of Congress Control Number: 2012947554

For all general information, please contact Arcadia Publishing:
Telephone 843-853-2070
Fax 843-853-0044
E-mail sales@arcadiapublishing.com
For customer service and orders:
Toll-Free 1-888-313-2665

Visit us on the Internet at www.arcadiapublishing.com

*To all those men and women whose dedication, creativity,
and hard work made Syracuse television history*

CONTENTS

ACKNOWLEDGMENTS

A book like this is only possible with the help of many people and organizations. First and foremost, we want to thank our partners in this endeavor: the Onondaga Historical Association (OHA), executive director Gregg Tripoli, and curator of history Dennis Connors. The archives of the OHA provided us a treasure of wonderful photographs and information, and we appreciate the work of research associate Sarah Kozma in helping us find and sort through the huge amount of information.

All three of us are employed by WSYR-TV NewsChannel 9, and we would like to take a moment to thank our boss, general manager Theresa Underwood, for making available the station's vast collection of photographs.

We also want to thank WCNY general manager Robert Daino, director of development Liz Ayers, and senior vice president John Duffy for free reign of the public station's hundreds of photographs, and to former staff members Richard Calagiovanni, Larry Wood, and Mike Clark for sharing the station's history. A big thank-you also goes to Granite Broadcasting and Matt Rosenfeld for their enthusiastic support of this project and for granting us permission to use station photographs.

YNN's Ron Lombard has always been a reliable supporter of professional organizations in the industry, and of the preservation of the history of local television, and he came through for us on this book.

We also need to thank the publisher of the *Syracuse New Times*, Bill Brod, and editor-in-chief Molly English-Bowers and photographer Michael Davis for being supportive of this project and allowing us access to their archives. Our thanks also go out to Klineberg Photography in Syracuse for permission to use a number of its wonderful photographs.

There are several former media employees who let us explore or use their personal photographs or shared with us their knowledge of our local television history. They include Mike Price and Jennifer Price, Ron Curtis Jr., Tony Rizzo, John Ellis, Dave Tinsch, Joe Gassner, Ron Gersbacher, John Nicholson, Al Fasoldt, Walt Sheppard, Caroline Stanistreet, Lorrie Conner, Jay Lurie, Mark Ogden, Shawn Milair Wayson, and Gary Hartman.

We also want to recognize our colleague at NewsChannel 9, Rod Wood, who has worked at both Channel 5 and Channel 9. He provided us with direction, insight, and help in identifying many of the people in our photograph collection. And as always, he kept us laughing.

INTRODUCTION

Most television viewers today take for granted the seemingly endless choices a click away from their fingertips. Who could have imagined, when the first television sets came on the market, that there would one day be whole networks devoted solely to food, fashion, music, news, and sports? But with all those options, there is one constant from the earliest days of television transmission: we still have local television stations broadcasting from here at home.

When television came to the Syracuse area in the late 1940s, the new medium was all about imagination and endless possibilities. Black-and-white images flickered on cathode-ray tubes, many of which were manufactured just north of Syracuse at the General Electric plant on Electronics Parkway in Liverpool.

Programming pioneers back then were believers. They believed they could turn ideas into programs; shows that would improve the quality of life of the people who watched them. In the early days, television sets were few and far between.

Whole families would get all dressed up and pile into the Nash Rambler for a trip downtown to the main shopping district along South Salina Street. They would ride the escalators at Dey Brothers or E.W. Edwards, the big department stores, to stand transfixed with other families in the electronics aisle. For hours, their faces would glow from a bluish light as they watched whatever might show up on one of the two channels the set-top antenna picked up.

A few years later, those same families would rearrange their living rooms in Eastwood or Bayberry to make room for a huge black-and-white console, an imposing piece of furniture. But a quick check of *Stars* magazine or *TV Guide* would send them scurrying across the street to watch the latest "Network Spectacular" on the neighborhood's first color television.

Every show back then was "Appointment Television." If you missed it when it first aired, the opportunity evaporated forever into the ether. The viewers of those days could not have dreamed of the changes to come over the course of 60 years. Local television today has gone digital, with images so vivid and lifelike on high-definition screens that one can detect a single blade of grass on a golf course or hear the chirp of crickets through the surround-sound speakers in home theaters. Favorite shows are available online and "on demand," so you can watch on your schedule, not the one set by Hollywood programmers. You do not even need an actual television set these days. Increasingly, people watch content on their laptops, iPads, or smart phones.

Syracuse's first television station, WHEN-TV, signed on the air on December 1, 1948. WSYR-TV followed in 1950. A dozen years later, Syracuse got a third station, WNYS. All three of these pioneering commercial stations have changed owners and call letters over the years, and they have been joined by more than a dozen other local channels and program services, including public broadcasting, cable, campus stations, and digital sub-channels.

As one old television jingle put it, the Syracuse television market stretches "from the Mohawk Valley to Ontario . . . from the Finger Lakes, there's a feeling you know." The national ratings services keep track of those feelings through polls that track how and when Central New Yorkers watch their favorite stations. Syracuse was once one of the top 50 markets in the nation. But, as population declined and towns like Utica and Watertown were spun off into their own markets, Syracuse has settled into the middle of the second 50 markets. Local broadcasters now consider their prime marketing area to stretch, roughly, north-to-south from Pulaski to Cortland and east-to-west from Rome to Waterloo. It includes all or part of eight counties and covers a little more than a million people, according to the latest census figures. Beyond that, stations will occasionally hear from viewers as far away as Kingston, Ontario; the Adirondacks; the Rochester area; and the Southern Tier. And streaming newscasts ride the Internet around the world, so snowbirds in Florida and exchange students in Europe can stay in touch with what is happening at home.

Technology has evolved from black-and-white to color, from monaural to stereo, from analog to digital, from static-filled blur to high-definition dream. But one thing remains the same: just as they have for more than 60 years, Central New Yorkers still turn to their favorite local stations for hometown news and the most comprehensive information available about their world, all delivered by familiar, friendly faces.

It can get a little confusing trying to sort out the history of Syracuse television. Channel 5 was originally Channel 8 on your dial, and Channel 3 started out on Channel 5. WSYR-TV was originally WNYS, while WSTM was originally WSYR. And WNYS is now another station altogether! Still with us? Here is a viewer's guide for Syracuse TV and this book:

WSTM is the local affiliate for the NBC television network on Channel 3. The station was originally called WSYR-TV and broadcast on Channel 5 for a short time when the station was launched.

WTVH is the CBS affiliate on Channel 5. It signed on as WHEN-TV Channel 8.

The current WSYR is the ABC affiliate and has broadcast on Channel 9 from the start, first as WNYS-TV and later as WIXT.

After Channel 9 dropped the WNYS call letters, the letters went to a Buffalo radio station for a time before returning with the launch of Channel 43 (MyTV).

The following stations are right where they have been since they first signed on: WCNY Channel 24 (PBS); WSYT Fox 68; and WSPX Channel 56 (now ION).

Of course, very few people watch television over the air anymore. Those with Time Warner Digital Cable, DirecTV, Verizon FiOS, or any other program service should disregard these channel assignments and check local listings to find the best local news and programming. To make things a little easier to follow, we have divided this book into chapters by station.

One

REMEMBER WHEN

Syracuse Television was born at 8:32 p.m. on December 1, 1948. The birthplace was a makeshift studio in an old industrial building that once produced components for bombsights used in World War II aircraft. According to newspaper accounts at the time, announcer William Bohen chatted with his guest, harpist Melville Clark, about his recent visit with England's Princess Elizabeth. Clark then entertained viewers with several tunes on the harp, including "Home on the Range." Those first steps into the new world of television were made possible by the frenetic work of employees of the Syracuse Meredith Television Corporation, which set a record by getting WHEN-TV Channel 8 on the air just 16 days after broadcasting equipment arrived on site.

The station would log television firsts for decades to come. Nine months later, the station moved outside the studio to broadcast several hours of programming from the New York State Fair. In 1982, the station began the tradition of broadcasting entire newscasts live from the fair.

In 1955, after extensive consultation with parents and educators, WHEN launched a new program aimed at preschoolers. *The Magic Toy Shop* was on the air for more than 27 years, with producer Jean Daugherty authoring 6,500 scripts. She is remembered best as "The Play Lady."

WHEN partnered with General Electric in 1964 to computerize election returns. In the 1976 election, 12 years later, the station was the first to use the roaming "Live Eye" to broadcast reactions from candidates at their party headquarters.

Kay Larsen helped busy housewives with recipes and other household problems. "Uncle Skip" and "Captain Joe" entertained the kids, and teenagers flocked to the *Top 10 Dance Party*.

Through the years, viewers saw Dick Grossman, Ray Owens, Art Goodwin, Ron Curtis, Rod Wood, Carrie Lazarus, Maureen Green, Matt Mulcahy, and Michael Benny deliver the news.

Hundreds of broadcast professionals got their start there, including *Today's* Al Roker, *Monday Night Football's* Mike Tirico, ABC News anchor David Muir, and CNN's Chris Lawrence and Bob Van Dillen.

Hundreds of people helped produce thousands of hours of programming in the station's more than 60-year history, creating countless memories. We regret that we can only touch upon a few.

The first home of Syracuse's first television station was an old factory building at 101 Court Street. It was the former Cine Simplex factory, which had made movie cameras and projectors. During World War II, components for the Norden bombsight used by American warplanes were produced there. (Courtesy Onondaga Historical Association.)

The first broadcasts began just 16 days after equipment arrived at the station, a new record in those early days of television. The station was owned by the Meredith-Syracuse Company. Meredith, based in Des Moines, Iowa, was the publisher of *Better Homes and Gardens* and *Successful Farming* magazines. The company owned the station until its sale to Granite Broadcasting in 1993. (Courtesy Onondaga Historical Association.)

Few people had television sets in 1949, so the station built interest and excitement in the new medium by bringing television productions to the public. Starting that year, WHEN-TV began an annual tradition of originating several hours of live programming late each summer from the New York State Fair. This is a photograph of the makeshift control room used to switch programs from cameras and audio sources at the fair. (Courtesy Onondaga Historical Association.)

This mid-1950s photograph shows a production of *The Magic Toy Shop* at the New York State Fair. On the menu board at left is the station's programming schedule from the fair, including weather, the religious broadcast *These Things We Share*, *Party Line*, *Gal Next Door*, and *Kay's Kitchen*. (Courtesy Onondaga Historical Association.)

Kay Larson was one of the pioneering women in Syracuse broadcasting. She hosted a daily cooking show called *Kay's Kitchen* weekdays at 9:30 a.m. beginning in 1952. Advertisers had yet to base their advertising purchases on demographics, but station management and sponsors both knew there was a large audience of housewives watching during the day. (Courtesy of the Onondaga Historical Association.)

Larson also hosted a show called *Gal Next Door*, which the station billed as covering topics and issues important to women. In this photograph, Larson and her guests discuss civil defense preparedness in the days when the Cold War and the atomic bomb were on everyone's mind. (Courtesy Onondaga Historical Association.)

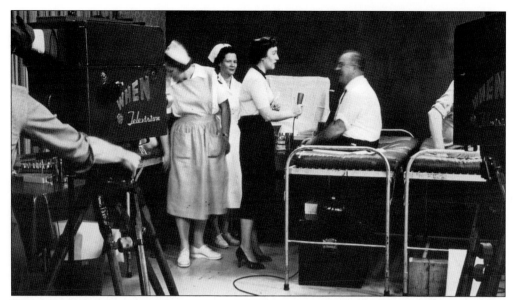

WHEN-TV often showcased its commitment to the community, with all kinds of special events. In this photograph, the station broadcast live from a blood drive being held in its studios to celebrate Channel 8's fifth birthday. Host Kay Larsen interviews a blood donor who dropped by the studio. (Courtesy Onondaga Historical Association.)

Events like the March of Dimes Telethon involved all hands, including news staffers. Here, the team has fun wearing hats with "press" cards in the bands and some very clichéd nicknames. Richard Grossman (far left) was WHEN-TV's first news anchorman. He attended Syracuse University Law School when he was not delivering the news. He went on to become one of the area's most prominent attorneys. (Courtesy Onondaga Historical Association.)

Ray Owens is at the anchor desk (left) and Tom Watkins works the scoreboard on November 6, 1956, as the WHEN-TV news team covers the election. Note the signs on the anchor desk and the wall with the name of the sponsor of that night's coverage, Marine Midland Bank. (Courtesy Onondaga Historical Association.)

Religious programming played a significant role in the early days. The station was the first in the nation to televise an entire Roman Catholic mass. In this 1955 photograph, WHEN-TV became the first station in the world to carry live coverage of the ordination of new priests. The station needed special permission from the Vatican for the broadcast. Note the big camera trying to stay inconspicuous. (Courtesy of Onondaga Historical Association.)

Notice the sign calling for silence in the balcony of Syracuse's Cathedral of the Immaculate Conception, which served as a makeshift control room and camera platform for the station's ordination coverage. Jean Daugherty was the director of the broadcast. (Courtesy of Onondaga Historical Association.)

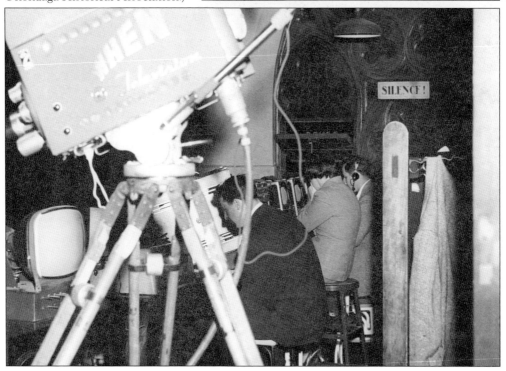

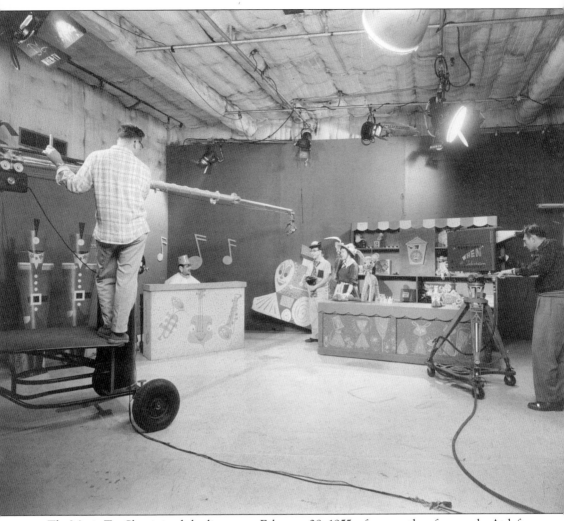

The Magic Toy Shop joined the lineup on February 28, 1955, after months of research. At left on the piano is Twinkle the Clown, played by station musical director Tony Riposo; at center is Eddie Flum Num, portrayed by staff artist Socrates Sampson; and at right is the Play Lady, show creator and producer Jean Daugherty. (Courtesy of Onondaga Historical Association.)

Marylin Hubbard Herr played the role of Merrily, the proprietress of *The Magic Toy Shop*. Each show opened with a reminder that "A smile is the magic key, to the magic door, to the wonderful Magic Toy Shop." Children were instructed to "put your thumbs at the corners of your mouth, like this, and turn them up, into a smile." (Courtesy Onondaga Historical Association.)

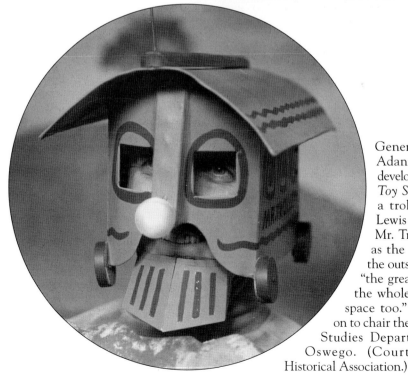

General manager Paul Adanti insisted during development of the *Magic Toy Shop* that there be a trolley in the show. Lewis O'Donnell played Mr. Trolley, who served as the children's link to the outside world by being "the greatest storyteller in the whole world and outer space too." O'Donnell went on to chair the Communications Studies Department at SUNY Oswego. (Courtesy Onondaga Historical Association.)

"Eddie Flum Num" (Socrates Sampson) threads the Flum-o-matic projector "that only shows movies boys and girls particularly like." Eddie dressed in a sort-of sailor suit, because he once worked on a showboat. He was an artist who drew illustrations to accompany Mr. Trolley's stories. (Courtesy Onondaga Historical Association.)

Below, creator Jean Daugherty performed the role of the Play Lady, a character originally created to fill in when Marylin Hubbard Herr ("Merrily") had a baby. The grandmotherly Play Lady and her Abracadabra Book took kids on adventures around the world, including trips to Washington, DC, the United Nations, and sometimes scary places like the doctor's office or the barbershop. (Courtesy Onondaga Historical Association.)

Jean Daugherty was the driving creative force behind *The Magic Toy Shop*. She wrote more than 16,000 television programs during her career, including every episode of *Toy Shop*. She also produced and directed episodes before taking on the character of the Play Lady. She served as the station's community affairs director and continued producing public affairs programming, holiday specials, and other community programs until retiring in 1995. (Courtesy *Syracuse New Times*.)

There was also a Saturday edition called *Toy Shop Jamboree*. Below, Eddie Flum Num (Socrates Sampson) hands out balloons to young fans. The cast made frequent public appearances to meet fans of the show. (Courtesy Onondaga Historical Association.)

In a segment of *Toy Shop Jamboree* called "Be a Clown," performers (from left to right) Lew O'Donnell, Tom Bader, Socrates Sampson, and Gordon Alderman showed viewers how clowns are created. Alderman was the station's program director and hosted or participated in many of the station's programs. (Courtesy Onondaga Historical Association.)

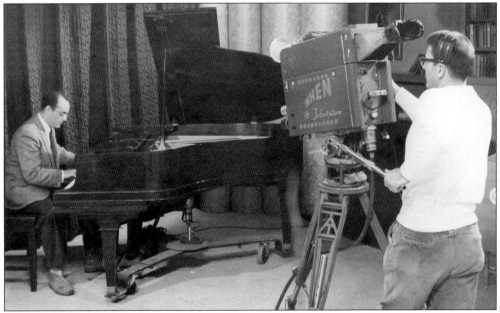

Tony Riposo served as WHEN-TV's music director. He is seen here performing in the 1950s. He also appeared as Twinkle, a clown who spoke only through music on the *Magic Toy Shop*. Riposo later made a name for himself as the music director for the popular McGuire Sisters in Las Vegas and on tour throughout the country. (Courtesy of Onondaga Historical Association.)

'CONGRATULATIONS MAGIC TOYSHOP'

he one and only cast of Magic Toyshop ready to celebrate their 5,000th sho

The show reached many milestones in its nearly three decades on the air. In this photograph for the station's newsletter, the cast of The Magic Toy Shop poses as it prepares to celebrate the broadcast's 5,000th episode. (Courtesy Onondaga Historical Association.)

John Scott, known on the air as Uncle Skip, hosted a talent show for local youngsters called *P&C Rascals*, sponsored by the P&C supermarket chain. (Courtesy Onondaga Historical Association.)

Today, broadcasters use Facebook, Twitter, and e-mail to interact with and build relationships with viewers. In its first two decades, television brought viewers into the studio to take part in programs and gave many people a chance to be on the air. Here, Julie and Jamie Grey sing on the *P&C Rascals* show. (Courtesy Onondaga Historical Association.)

Debuting on April 11, 1958, *Fore* was a 15-minute program that featured instructions and tips on the game of golf. Area golf professional Bucky Hewitt of the Westvale Country Club and WHEN-TV's Arnie D'Angelo conducted the lessons. D'Angelo learned the fundamentals of the game on the air as viewers followed along. Drawings and a jackpot of prizes were featured on the premier of the program. (Courtesy Onondaga Historical Association.)

From the beginning, live fashion shows were a regular feature of the WHEN lineup, sponsored by local department stores. This scene is from a fashions luncheon in the spring of 1958. The broadcast originated at the Persian Terrace in the Hotel Syracuse and was hosted by the Addis Company. (Courtesy Onondaga Historical Association.)

Art Peterson, at the piano, welcomes viewers to *Club Continental* in this April 1954 photograph. The program, which aired from 6:30 p.m. to 6:45 p.m. on Monday, Wednesday, and Friday evenings, sought to give viewers the illusion that they were actually attending a nightclub. Peterson introduced film clips of various musical performances. That illusion apparently worked; viewers would often call asking for a reservation to the club, thinking it was real. (Courtesy Onondaga Historical Association.)

Below, singing sensation Johnny Mathis (left) shares a beverage provided by *Top 10 Dance Party* sponsor Coca-Cola. Mathis appeared on one of the *Dance Party* shows broadcast in August 1957 from the New York State Fair, where he performed his hit single "Wonderful, Wonderful." (Courtesy Onondaga Historical Association.)

The *Top 10 Dance Party* was a popular show with teenagers. Hundreds of young people from around Central New York flocked to the New York State Fair, where the station conducted dance contests throughout the run of the fair, carrying it all live. (Courtesy Onondaga Historical Association.)

Many changes took place in the 1960s. As seen in the outfield billboard at Syracuse's MacArthur Stadium, WHEN-TV was now Channel 5. The Federal Communications Commission reallocated television channels across upstate New York in 1962 in order to accommodate more stations. The move shifted WHEN from channel 8 to channel 5 and opened the way for Channel 9 to sign on. (Courtesy Onondaga Historical Association.)

Parent company Meredith Corporation built a new facility to accommodate WHEN's television and radio production needs. The move to the new building was completed by early 1964. The building was located at 980 James Street. The James Street corridor was coming into its own as a prime location for commercial and apartment development. (Courtesy Onondaga Historical Association.)

The new channel assignment, new building, and new decade did not affect WHEN-TV's focus on local programming that served the community. Members of the Syracuse Parochial School concert band perform in May 1963 on *One O'Clock Scholar*, a program that highlighted the talents of schoolchildren from across Central New York. (Courtesy Onondaga Historical Association.)

In this May 1964 photograph, program director and host Gordon Alderman talks with students from the Jamesville DeWitt school district about their art projects. Students from as far away as Utica and Penn Yan took part in *One O'Clock Scholar*, as the station cast a wide net for young talent. (Courtesy Onondaga Historical Association.)

If the studio was not big enough to accommodate the performers, the crew just cleared the parking lot and rolled the cameras out the back door. This June 1965 photograph shows members of the Gillette Blue Blades marching band from the North Syracuse School District. (Courtesy of Onondaga Historical Association.)

CHANNEL 5 *JET SET* weekdays at 4:30 in *color*

The 1960s also ushered in color television to Central New York. This promotional flyer features the kids' show *Jet Set.* Live-action hosts Captain Skip (John Scott, left) and Captain Joe (Joe Ondrick, right) hosted *Roger Ramjet* cartoons on weekday afternoons, capitalizing on the excitement of the "Space Age." The station debuted its color programming on election night 1966. (Courtesy Onondaga Historical Association.)

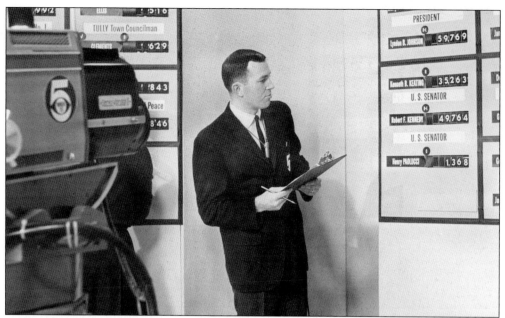

Election coverage took a big step into the modern era in 1964, when the station partnered with General Electric to computerize election returns. In this photograph, WHEN's Brian Madden is at one of the tote boards reporting on local results in the presidential race between Lyndon Johnson and Barry Goldwater and in the US Senate race between Kenneth Keating and Robert Kennedy. (Courtesy Onondaga Historical Association.)

Results were called in to people on phones at GE's massive Electronics Park factory and research complex in the town of Salina. The information was entered onto punch cards for processing by GE's computers and then sent over phone lines and intercoms to people operating the tote boards in the studios. (Courtesy Onondaga Historical Association.)

Ron Curtis joined the station in 1959 after years in local radio. His first major assignment on the television side was as the "Atlantic Weatherman," wearing the uniform of the sponsor, Atlantic Richfield gasoline. Curtis told interviewers years later that he resisted the role, but was told in no uncertain terms by management that if he wanted a paycheck, he would do it. (Courtesy of Onondaga Historical Association.)

In 1966, Curtis was appointed anchor of the Channel 5 Report at 11 O'Clock weekday evenings. Ron later anchored the news broadcasts at noon, five, and six. Ron had the ability to easily edit news copy on the fly, often ad-libbing and greatly improving on the news copy written by a reporter or producer. (Courtesy Klineberg Photography.)

Ron Curtis was often called the Walter Cronkite of Syracuse for his trustworthiness, class, and calm demeanor in front of the camera no matter the chaos going on around him. He spent more than 41 years with channel 5, thirty-six of them at the anchor desk. In this photograph from the late 1970s—before personal computers—Curtis is seen ripping state and national news from the wire service Teletype machines. (Courtesy *Syracuse New Times*.)

Today's Al Roker got his first job in television at WHEN-TV while he was still a student at the State University of New York at Oswego. He started as a weekend weatherman in 1974 and was promoted to the weekday job the next year. In this photograph, Roker is seen doing artwork for the station. Among his creations was a caricature of himself used in a Fahrenheit to Celsius conversion chart mailed to viewers. (Courtesy *Syracuse New Times*.)

This photograph shows the news set in use during the early 1970s. Pictured are anchor Art Goodwin (left) and sportscaster Jack Morse. Morse produced a feature called "Athlete of the Week" that profiled local scholastic sports stars. It was a popular feature he brought with him when he moved to channel 9 in the mid 1980s. (Courtesy *Syracuse New Times*.)

Art Goodwin was an integral part of the channel 5 news operation. He served as news director, anchored news broadcasts, hosted public affairs shows like *Dialogue* and *Camera 5*, and contributed his producing expertise to local production of the infotainment program *PM Magazine*. (Courtesy *Syracuse New Times*.)

In April 1983, WTVH was one of about 40 television stations honored by Pres. Ronald Reagan for broadcasting job fairs in an effort to help the unemployed during the recession. Pictured outside the White House are, from left to right, Lou Gulino, who produced the job fair broadcast; Ron Curtis, who anchored; and photographer Dave Tinsch, who covered the event. (Courtesy Dave Tinsch.)

On election night 1982, *Newscenter 5*'s Joe Sottile (center) is at Democratic Party headquarters preparing for a live report. Election night coverage was always an intense competition between local stations wanting to be the first to post results and get interviews with the winning candidates. (Courtesy *Syracuse New Times*.)

Like many staffers at the station, Jack Slater played multiple roles; he was a director, an advertising pitchman, a weathercaster, and a host of *PM Magazine*. He is seen here in the late 1970s standing in front of the Live Eye, the first mobile news truck in Syracuse. (Courtesy of *Syracuse New Times*.)

Beginning in 1982, Newscenter 5 began broadcasting its entire 12:00 p.m. and 6:00 p.m. newscasts from the New York State Fair. The station constructed the weatherproof box studio below in 1983, adjacent to the Pepsi International Building, where 34 years earlier, the station had broadcast hours of live entertainment programs. Seen here, from left to right, are cameraman Don Miller, anchors Ron Curtis and Maureen Green, and cameraman John Fear. (Courtesy Lorrie Conner.)

In this photograph, producer Robin Lisi (left) confers with producer Lou Gulino (center) and director Dominick Cogliandro (right) in advance of a news broadcast during March 1993 as a blizzard blanketed the region. Though computers were used to write and produce newscasts, printed scripts were spread on a table so the director could mark cues for camera changes, audio commands, and insertion of videotape. (Courtesy Joe Gassner.)

Tracy Davidson (left) prepares to anchor one of many newscasts and special reports during the blizzard that dumped more than three feet of snow on the area. Meteorologist Kathy Orr (right) put in a long weekend of weather updates. Both women went on to long careers in the Philadelphia market. (Courtesy Joe Gassner.)

Scott Atkinson was one of the best television reporters in Syracuse during the 1980s and 1990s. Here, he is preparing to go live with photographer Al Nall outside the station during coverage of the blizzard of 1993. Scott is now news director at WWNY-TV in Watertown, New York. (Courtesy Joe Gassner.)

The anchor team at Newscenter 5 in the mid-1980s included, from left to right (first row) Carrie Lazarus, Bob Kirk, news director Jim Holland, Ron Curtis, and Maureen Green; (second row) Joe Sottile, Jack Slater, Rich Isome, and Dennis Stauffer. During this period, Newscenter 5 was the dominant station in the ratings for all newscasts. (Courtesy Klineherg Photography.)

In 1989, the Newscenter 5 team included, from left to right, (first row) Lisa Holbrook, Maureen Green, Ron Curtis, Tracy Davidson, and Keith Kobland; (second row) Kathy Orr, John Fisher, news director Graham Robertson, Mike Tirico, and Liz Ayers. Tirico began doing weekend sports at the station while still a student at Syracuse University. Today, he is the host of Monday Night Football on ESPN. (Courtesy Klineberg Photography.)

David Muir joined WTVH as a reporter and anchor right after graduating Ithaca College in 1995. But the Syracuse-area native was a part of the team long before that. He contacted the station as a youth asking for a tour and career advice. He spent school breaks in high school and college as an intern. Muir left Syracuse in 2000 for WCVB in Boston and joined ABC News in 2003. He is now a weekend anchor at the network, Diane Sawyer's primary backup on the anchor desk, has covered stories around the world, and is cohost of the news magazine *20/20*. (Courtesy Klineberg Photography.)

In this photograph from 1993, Tony Rizzo (left) is running the assignment desk at WTVH. His job is to coordinate all the movements of reporters and photographers, keep track of stories, and react to breaking news. Tony is now an IT professional at WUSA. Bill Carey (right) was news director. Bill came to WTVH after many years running the radio newsroom of WHEN, and was the creative force behind many of the station's highly acclaimed specials. He is a field anchor and senior reporter at YNN. (Courtesy Joe Gassner.)

Ron Curtis and Maureen Green are seen here in the 1990s. Curtis was inducted into the New York chapter of the National Academy of Television Arts and Sciences Silver Circle. He and Green anchored the only newscast in Syracuse to receive a New York Emmy. Ron retired on December 1, 2000, and died the following year. (Courtesy Klineberg Photography.)

The anchor team in the early 2000s featured, from left to right, sports director Kevin Maher, anchors Matt Mulcahy and Maureen Green, and chief meteorologist Tom Hauff. (Courtesy Klineberg Photography.)

This is the last anchor team to operate out of the facilities at 980 James Street. In April 2009, most of the staff was let go when ownership contracted with WSTM/NBC 3 to operate the station and produce newscasts. From left to right are Keith Kobland, Michael Benny, Kevin Maher, Donna Adamo, Tom Hauff, and John DiPasquale. (Courtesy Klineberg Photography.)

Two

CURLY'S KINGDOM

WSYR-TV was the second television station to hit the airwaves in Syracuse, on February 15, 1950. It became the city's NBC affiliate and went on the air as Channel 5—WHEN was still on Channel 8 at the time. Before long, the Federal Communications Commission shifted WSYR to Channel 3 on the dial.

The station's first owners were no strangers to communications. The Newhouse family already owned Syracuse's two daily newspapers and WSYR Radio. WSYR Radio executive E.R. "Curly" Vadeboncoeur was put in charge of the new enterprise. His presence was felt behind the scenes and on the air, as he turned the station into a popular destination for television viewers throughout the Syracuse area.

The station gave its top radio stars their own television shows in a lineup that included everything from farm shows to cooking shows to news broadcasts. Newly minted television personalities like Ed Murphy and Jim Deline were already household names in Central New York.

WSYR-TV's original studio was located in a former auto repair garage on Harrison Place, right next to the Kemper Building. The station also used Syracuse University's television studio, in the basement of Carnegie Library on the main quad. In 1958, Channel 3 opened a state-of-the-art broadcast center for both radio and television at 1030 James Street. The station still occupies that building today.

Early institutions included Fred Hillegas, who anchored the news from the station's earliest years until 1972; Joel Mareiniss, a newsman-turned-sports director who served as the "Voice of the Orange," covering Syracuse University for many years; and Kay Russell, the host of the long-running morning show *Ladies Day*.

In 1980, the Newhouse family sold WSYR-TV to the Times Mirror Company, which changed the call letters to WSTM-TV. Over the next 30 years, the station had several owners. In 2009, current owner Barrington Broadcasting entered into an agreement with WTVH owner Granite Broadcasting in which Barrington would operate WTVH and produce news for WTVH. Both stations now operate out of the WSTM facilities at 1030 James Street.

In the 1950s, WSYR's radio and television journalists were given marked station wagons to gather news and interviews for the daily newscasts. The news vehicle is seen here parked outside of the Kemper Building in downtown Syracuse, where the station had its control room and a small studio. Another studio was located on the Syracuse University campus. (Courtesy Onondaga Historical Association.)

WSYR-TV was fast outgrowing its first makeshift studios. By the late 1950s, plans were in the works for a new broadcast facility. Ground was broken in the 1000 block of James Street for a facility that would house offices and studios for WSYR radio and television, all under one roof. (Courtesy Onondaga Historical Association.)

When the new facilities were completed in June 1958, television viewers stood in a long line for a chance to be among the first to see the new state-of-the-art broadcast center at 1030 James Street. (Courtesy of Onondaga Historical Association.)

E.R. "Curly" Vadeboncoeur was known far and wide for his battlefield reporting from World War II for WSYR Radio. He helped put the television station on the air, and it operated under his direction until 1980. Vadeboncoeur eventually rose to the presidency of Newhouse Broadcasting. In this photograph, Vadeboncoeur displays his "short-snorter." It was a practice of some airmen and troops in World War II to collect currency, tape the notes together, and obtain signatures of friends. (Courtesy Onondaga Historical Association.)

Special projects director Don Edwards (above, left) and Curly Vadeboncoeur are seen on an election night in the early 1970s. Though he was a broadcasting executive for a long time, Vadeboncoeur remained a newsman at heart. His regular election night appearances were introduced with a salute to "the dean of Syracuse newsmen." (Courtesy Onondaga Historical Association.)

These RCA color cameras in the WSYR-TV studios were even bigger than the original black-and-white cameras. The station's network, NBC, led the way in providing color programming, promoting it with its famous colorful peacock logo. (Courtesy Onondaga Historical Association.)

Jim Deline and the Gang gained a huge following with their popular midday music and variety show. It featured a studio orchestra comprised of some of the area's finest musicians, including Leighton "Sox" Tiffault, Carl Mano, and Myron Levee. (Courtesy Onondaga Historical Association.)

Jim Deline (at the microphone) joined WSYR Radio in 1950 from WFBL. He moved his variety format to the noon hour at Channel 3 to great success. Deline showcased local talent from around Central New York as well as entertainers passing through town until his death in 1963. (Courtesy Onondaga Historical Association.)

Denny Sullivan hosted the *Popeye Funhouse* in the early 1960s. He dressed as a carnival barker, wearing his signature red-striped coat and twirling a cane. The late-afternoon show gave the kids a diversion while mom cooked dinner. (Courtesy Onondaga Historical Association.)

Below, Denny Sullivan welcomes faithful viewers to the set of the *Popeye Funhouse*. A popular feature of the show was an early form of interactive television, when Denny would show photographs of young viewers to celebrate their birthdays. (Courtesy of Onondaga Historical Association.)

47

After Jim Deline died, Channel 3 turned Deline's midday show over to Denny Sullivan. Here, Sullivan appears on the set of his show with NBC Saturday morning personality Shari Lewis on August 6, 1963. (Courtesy Onondaga Historical Association.)

The legendary No. 44, Syracuse football great Jim Brown (center), headlined the WSYR Appreciation Dinner in downtown Syracuse on January 2, 1958. Brown had gone on to stardom with the NFL's Cleveland Browns. He is flanked by WSYR Channel 3 general manager E.R. "Curly" Vadeboncouer and sports director Bill O'Donnell. In 1966, O'Donnell left Syracuse to join the Baltimore Orioles and later became part of NBC's *Baseball Game of the Week.* (Onondaga Historical Association.)

Ed Murphy's deep voice and smooth, easy broadcast style made him a household name in Central New York. His radio audience followed him into television in the 1960s, when he hosted an afternoon film series called *Ed Murphy's Hollywood Matinee.* During this period, he was also *The Time Keeper,* hosting the morning-drive radio show on WSYR-AM. (Courtesy Onondaga Historical Association.)

One of the most popular features of *Ed Murphy's Hollywood Matinee* was the daily "Dialing for Dollars" contest, where Murphy would dial numbers drawn from the pages of the phone book and ask viewers if they knew "the count and the amount." This winner takes home a $450 prize just in time for Christmas on December 17, 1969. (Courtesy Onondaga Historical Association.)

Kay Russell hosted *Ladies Day*, with daily broadcasts on both television and radio. The morning television program became the longest-running women's show in the country. (Courtesy Onondaga Historical Association.)

Kay Russell's show often offered viewers sound advice on how to stay safe at home, like this swimming pool demonstration, broadcast from just outside the James Street studios. (Courtesy Onondaga Historical Association.)

Along with cooking and lifestyle segments, *Ladies Day* regularly featured the latest in fashions, provided through the courtesy of Syracuse's locally owned department stores. Here, Kay Russell checks out the latest fashions in 1970. (Courtesy Onondaga Historical Association.)

Kay Russell got to ride around town in style. Here, she is ready to make a public appearance in downtown Syracuse in a highway courtesy car provided by Paul T. Henson Motors, "the Friendly Ford Dealer." (Courtesy Onondaga Historical Association.)

In this photograph, Alan Milair looks like a lot of television hosts of the era. He was a familiar face on WSYR-TV and spent more than 30 years on WSYR radio. But viewers never saw his face when he portrayed one of the station's most popular characters, Dr. E. Nick Witty, the host of *Monster Movie Matinee*. (Courtesy Onondaga Historical Association.)

Milair had intended to let viewers determine what Dr. Witty might look like. But the show took off quickly, and Milair feared his low-budget attempts at makeup would never beat the viewers' imagination. So Dr. Witty never revealed his face during the program's 16 years, welcoming his "Dear Guests" for a visit with his trusted aide, Epal, played by Bill Everett. In this publicity photograph, Milair's face is etched out. (Courtesy Onondaga Historical Association.)

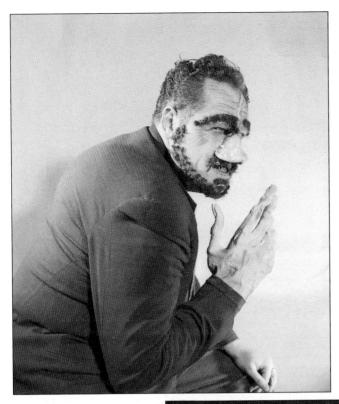

Epal sounded like a fitting name for Bill Everett's lab assistant and companion to Dr. Witty on *Monster Movie Matinee*—something right out of a horror movie. In fact, the name came right off his birth certificate. Everett was born Willard Everett Lape Jr., and he just reversed the letters of his last name to spell "Epal." (Courtesy Onondaga Historical Association.)

Bill Everett became Salty Sam on Channel 3's Saturday morning *Popeye* show in the early 1960s, when host George Wilson left the station. Years later, East Syracuse native Tom Kenny told people he modeled Patchy the Pirate on *SpongeBob SquarePants* after the guy in his hometown who "hosted the kids show, filled in on *Bowling for Dollars* and did the weekend weather." (Courtesy Onondaga Historical Association.)

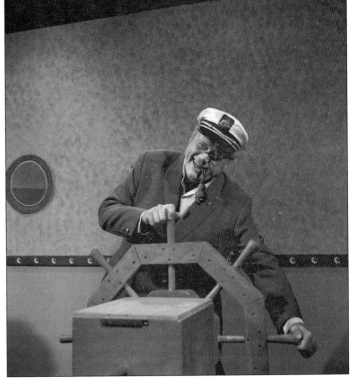

Popeye the Sailorman was in good hands in the Syracuse market in the early 1960s. Salty Sam hosted a Saturday morning *Popeye Theater*, and Denny Sullivan was in charge of the *Popeye Funhouse* on weekday afternoons. (Courtesy Onondaga Historical Association.)

Bob "Deacon" Doubleday, famous for his signature line, "This is the Deacon speakin'," spoke to the farm crowd "From the Wired Woodshed" on his morning radio show and in his appearances on WSYR-TV. Here, he does a cooking demonstration at the Home Show in 1959. (Courtesy Onondaga Historical Association.)

Claude "Red" Parton (above) delivered a brief local sportscast that followed 10 minutes of news in the evening during WSYR-TV Channel 3's early days. Note the advertisements for local brewer Utica Club beer on the back wall. The longtime sportscaster's career spanned more than six decades. He was the first journalist inducted into the Syracuse Sports Hall of Fame. (Courtesy Onondaga Historical Association.)

Sports announcer Joel Mareiniss holds a baseball bat for a publicity photograph in the 1960s. He was better known as the "Voice of the Orange," doing radio and television broadcasts for Syracuse University football and basketball games. (Courtesy Onondaga Historical Association.)

Channel 3 personalities would often make appearances to promote local sponsors. Joel Mareiniss is seen here at the opening of an Esso gas station on Onondaga Boulevard in Syracuse in 1959. The station offered a grand opening deal on tires, selling them for just $12.95 each. (Courtesy Onondaga Historical Association.)

Dick Hoffman was a longtime newsman with WSYR-TV and WSYR Radio, serving as a reporter, anchorman, and news director. He is seen here on the news set, delivering a broadcast sponsored by D'Jimas Furs. Sponsorship placards like this were common until the mid-1960s. (Courtesy Onondaga Historical Association.)

John Frederic "Fred" Hillegas began his career as a reporter for the *Post-Standard* in 1938. He joined WSYR radio in 1946 as a newscaster and added television news broadcasts to his duties when WSYR-TV signed on. He remained a fixture in local news until he retired in 1972. News sets in the early days were a simple affair, with just a desk, a sponsor logo, and a world map. (Courtesy Onondaga Historical Association.)

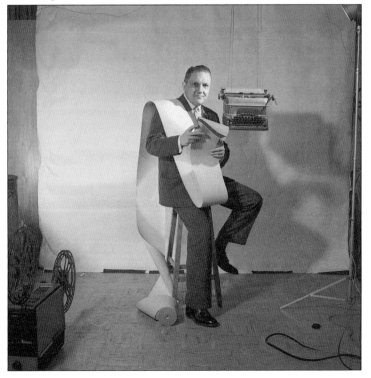

The WSYR promotions team had a little fun with this shot from the 1950s. They posed Fred Hillegas on a stool with wire copy over his shoulder, a typewriter hanging in front of him, and a film projector on the floor. Local news broadcasts were only 10 minutes long in the early days; they did not expand to 30 minutes until well into the 1960s. (Courtesy Onondaga Historical Association.)

This is a newsroom photograph of the WSYR-TV3 "Total News Crew" from the mid-1970s. From left to right, news anchors John Banks, Ron Hastings, and Jerry Barsha are in front, with sports director Joel Mareiniss and weatherman Bud Hedinger behind them. Barsha was the station's first weekend anchor and broke many exclusive stories. In his memory, the Jerry Barsha Memorial Scholarship helps students pursuing a career in broadcasting at Onondaga Community College. (Courtesy Onondaga Historical Association.)

Syracuse University graduate Laura Hand joined the Channel 3 newsroom in 1972. She spent much of that time as a reporter, a producer, and the anchor of the early-morning and noon reports. She was honored with a plaque on the Syracuse Press Club's Wall of Distinction in 2001, in part for her strong support of community causes. (Courtesy Onondaga Historical Association.)

The TV3 news team is on the run in this promotional shot taken outside the front entrance of 1030 James Street. They are, from left to right, Jerry Barsha, Dick Hoffman, Paul Ennis, Steve Kroft, Don Edwards, Johnny Bowles, Dave Northrup, John Nicholson, Laura Hand, John Banks, and Ron Hastings. (Courtesy Onondaga Historical Association.)

Channel 3 has served as a training ground for many young journalists. Steve Kroft worked on James Street after he graduated from Syracuse University's Newhouse School of Public Communications. He is seen here in 1970, before moving on to CBS News and 60 Minutes. (Courtesy Onondaga Historical Association.)

Investigative reporter Jim Kenyon spent a short time at WNYS Channel 9 before signing on with Channel 3 in 1976. He is a graduate of the journalism program at SUNY Morrisville, where a scholarship is given in his name each year. Kenyon has won numerous local and statewide awards for his reporting. (Courtesy Onondaga Historical Association.)

Known as "Consumer Man," Herb Weisbaum was part of the TV3 Total News Crew in the 1970s. He is an Emmy Award-winning reporter who has exposed everything from quack medicine to bogus investments. He began his career at Channel 3 in 1974 as an investigative reporter. After moving to Seattle, his consumer reports often appeared on NBC's *Today*. (Courtesy Onondaga Historical Association.)

Paul Ennis was just a sophomore at Syracuse University when WSYR hired him as a weekend reporter. Paul eventually became the station's assignment editor and producer of its 6:00 p.m. evening news broadcast. He was hired away by channel 9 as its executive producer. He won many awards and is credited with helping NewsChannel 9 solidify its number one standing in local news ratings. (Courtesy Onondaga Historical Association.)

Jackie Robinson went from newsroom secretary to anchor in her 34-year career at WSTM, making history in the process. When the State Regents exam asked eighth-graders to write about "Jackie Robinson," they meant the baseball Hall of Famer. But local students wrote about the news anchor, and the state accepted their essays. She became a New York State Broadcasters Hall of Famer herself in 2012. (Courtesy *Syracuse New Times*.)

Saturday Showboat was the last in a long line of WSTM-TV3 children's shows. Parents could trace its roots back to their own childhoods spent with "Salty Sam" in the early 1960s. By 1989, the show was still afloat, with magician George Becker and "Crafty Lady" Rosanne Caporuscio entertaining a studio full of kids. "The Showboat" cruised on until 1998. (Courtesy *Syracuse New Times*.)

Three

Colorful Channel 9

By 1960, Syracuse was the biggest market in the country still served by just two television stations. The Federal Communications Commission approved a third license in 1961, unleashing a battle over who would own the new station. At least nine owner groups wanted it. Rather than wait any longer, the FCC forced an unprecedented solution: those potential owners had to work together to put the new station on the air. A tower was erected in Pompey, south of Syracuse. Studio space was carved out of a vacant furniture store at Shoppingtown DeWitt. And on September 9, 1962, WNYS Channel 9 signed on the air. For the first time, ABC had a full-time home in Central New York.

The competition had a big head start, so "Colorful Channel 9" decided to do things differently. They brought a young, fresh approach to television, loading the schedule with live programming and interaction with the public. Early hits included *The Ben Schwartzwalder Show*, featuring the legendary Syracuse University football coach; *The Market Place*, originally an afternoon chat show; and *Hollywood Special*, late-night movies on the weekend. A young staff announcer, Mike Price, introduced the films with his best Bela Legosi impersonation, and "Baron Daemon" was born. For five years, his goofy antics were Channel 9's most popular feature, first on late night and later on an afternoon kiddy show.

That young, fresh approach has guided the station throughout its history. When local news really hit its stride in the 1970s, Channel 9 started reaching out to recruit established names, from Rod Wood, Bud Hedinger, and Dave Cohen to Carrie Lazarus, Doug Logan, Dave Eichorn, and Dan Cummings. Along the way, homegrown talent like Christie Casciano, Dave Longley, and Steve Infanti were added to the mix. The station became first a contender and then a leader in local news ratings.

Over the years, Channel 9's call letters have changed from WNYS to WIXT to WSYR. The news brand has evolved from 9 News to 9 Eyewitness News to NewsChannel 9. But Channel 9 in Syracuse still stands for a fresh approach and a strong commitment to the community.

Channel 9's earliest news format, *Information 9*, featured anchorman Martin Ross and sports director Carl Eilenberg. The original 1962 newscasts were just 15 minutes in length, at 6:00 p.m. and 11:00 p.m. Ross later anchored the news at WJW in Cleveland. He passed away in 1973. Eilenberg was the public address voice of the Carrier Dome for more than 20 years and served as mayor of nearby Rome, New York. (Courtesy NewsChannel 9.)

One of Colorful Channel 9's earliest shows was a regular Saturday morning visit to the *Bar J Ranch*. "Cowboy Bob" Hart (left) had an advantage over many children's television hosts: he really was a cowboy. He came to television from his mother's upstate horse-breeding farm. Later, he would serve as a ranch hand at a Christian children's camp in rural Maryland. (Courtesy NewsChannel 9.)

Channel 9 was the Syracuse home of *Romper Room*, the television nursery school. "Miss Cathy" Stampalia was the first teacher, beginning in 1962. She later created and hosted *Ladybug's Garden*, produced in its first year at Channel 9. Catherine Stampalia Hawkins later developed the TV/Radio Department at Onondaga Community College, which dedicated its TV studio to her when she retired after 25 years. (Courtesy NewsChannel 9.)

"There's a time and a place for everything. It's 11 o'clock. It's . . . *The Markert Place!*" Phil Markert came home to Syracuse when Channel 9 signed on in 1962. His was the first face seen when the station signed on. *The Markert Place* was a live talk and entertainment show that aired in the afternoons in the early 1960s. It later returned as a late-morning show in the early 1970s. (Courtesy NewsChannel 9.)

WNDR DJ Bud Ballou took over Channel 9's version of *American Bandstand* in 1965. Rolland Smith had launched *Dance Party* before anchoring at WCBS-TV in New York and then, for a short time, hosting *The Morning Program* on CBS. Ballou spent a little more than a year at Channel 9 before moving to Denver and later Boston. Among his special guests on the renamed *Bud Ballou Show* were the Shangri-Las, who sang the hit "Leader of the Pack." (Courtesy Ron Wray.)

In the early 1960s, Colorful Channel 9 made room in its children's lineup for holiday visits from Santa Claus. The Jolly Old Elf, who sounded a lot like staff announcer Charlie Featherstone, read letters each afternoon that had been sent by kids from across Central New York. (Courtesy NewsChannel 9.)

Mike Price's "Clown in a Cape" launched around Halloween 1962 and became a monster hit. "Baron Daemon" originally appeared in funny sketches that introduced horror films on the late night *Hollywood Special* on Saturday nights. When parents complained that their children would not go to bed until they saw the Baron, Channel 9 gave an afternoon kids show to *The Baron and His Buddies*, complete with a studio audience. (Courtesy NewsChannel 9.)

Cub Scouts and neighborhood groups were regular guests on *The Baron and His Buddies*. The Baron himself, Mike Price, is seen here at the helm of his spaceship. It took a little imagination for kids to "blast off" with the Baron. Some kids got into it more than others. (Courtesy Mike Price.)

This is a special moment for the Baron (Mike Price). He and Very Harry (Dennis Calkins) pose for a fourth-birthday photograph with Price's daughter Jennifer. Very Harry gave new meaning to the term "sidekick"—any time they ran out of scripted material, the Baron would kick him in the side, and then they could fill another five minutes rolling around. (Courtesy Jennifer Price.)

Baron Daemon recorded what is still considered the best-selling song in Syracuse history in 1964, "The Transylvania Twist." It was produced by Mike Riposo at his studios in the old Hotel Onondaga. Sam and the Twisters adapted their hit "Fooba Wooba John" for the melody, with lyrics by Hovey Larrison. The Bigtree Sisters added background vocals. The song has been "resurrected" by local radio stations every Halloween since. (Courtesy Mike Price.)

The Baron first appeared around Halloween 1962, hosting late-night films. His funny sketches quickly attracted an audience, including little kids who wanted to stay up way past their bedtime. When parents complained, station management ordered up more Baron, five days a week in the afternoon. The show ended when a studio fire destroyed the props and costumes in 1967. (Courtesy Mike Price.)

From his debut in 1962, Baron Daemon was constantly on the move around upstate New York. Mike Price made appearances as far away as Geneva, Old Forge, Cortland, and Watertown. He is seen here scaring up fans at the Utica-Rome Speedway in 1965. (Courtesy *Gater Racing Photo News.*)

Baron Daemon was not Mike Price's only on-screen character. For a time in the 1960s, he also hosted Warner Bros. cartoons on a morning children's show as "Cousin Orkie." The main set decoration was a bale of hay to sit on. Price admits his hayseed voice was influenced by Edgar Bergen's "Mortimer Snerd." (Courtesy NewsChannel 9.)

The Madisons appeared on the *March of Dimes Celebrity Parade* telethon from Channel 9's Shoppingtown studios in 1965. Syracuse television and radio stations traditionally put competition aside and came together to raise money for the fight against polio and birth defects. (Courtesy Ron Wray.)

Sports director Claude "Red" Parton joined Channel 9 in the mid-1960s after calling Syracuse University basketball and Syracuse Chiefs baseball games and handling public relations for the NBA's Syracuse Nationals. Among his duties at 9, Parton hosted SU's legendary football coach on *The Ben Schwartzwalder Show*, an autumn staple that started during the station's first weeks on the air. Parton's career lasted into the 1980s and spanned six decades. (Courtesy NewsChannel 9.)

WNYS Channel 9's promotions staff reviews an advertising layout promoting the *David Frost Show*, a syndicated 90-minute talk show that ran from 1969 to 1972. Graphic artists George Folsom and Dennis Calkins flank promotion director Carole Schell in Channel 9's art department in Shoppingtown. All three were original employees of the station. (Courtesy NewsChannel 9.)

Retired Air Force major Horace Meredith was the first certified meteorologist on Syracuse television. Viewers first heard "Meteorologist Meredith's" North Carolina drawl as he called in reports during the historic blizzard of 1966. He became well known for his maps on the Weather Cube, and his "Be Weatherwise" educational tips. "Stormy" Meredith was also one of the area's finest senior golfers. (Courtesy NewsChannel 9.)

Investigator Rod Carr was seen across the dial as the Syracuse Police spokesman for years, and later served as the police chief of the village of Phoenix, New York. But in the late 1960s, he served for a short time as a news anchorman at WNYS-TV Channel 9. (Courtesy NewsChannel 9.)

The Markert Place returned in the early 1970s as a morning show. Phil Markert insisted that guests appear live in the studio, which made for early wake-up calls for late-night performers like Mickey Rooney, Al Martino, and Connie Francis. Below, bandleader Lawrence Welk and his clarinetist, Syracuse native Peanuts Hucko, jam with the studio band. Markert is on piano. (Courtesy NewsChannel 9.)

Teams from high schools around Central New York tested their knowledge on *It's Academic!*, Channel 9's weekly quiz show. It was produced locally, using a nationally syndicated format from the same company that created *Romper Room*. Cicero High School took on Phoenix Central on this episode, moderated by Phil Markert. Channel 9's Carl Eilenberg and radio personality Bill Merchant also hosted the show during its Syracuse run. (Courtesy Phil Markert.)

Channel 9 picked up *Bozo, The World's Most Famous Clown* in the late 1960s. Similar to *Romper Room* and *It's Academic!*, the station produced its own live-action version of the show each weekday afternoon at 4:00 p.m., using costumes, wigs, and animated cartoons provided by the copyright holder—in this case Larry Harmon Productions. Mike Lattiff brought Bozo to life under the Shoppingtown "Big Top." (Courtesy NewsChannel 9.)

Channel 9 joined the Love Network in 1971, soon after Jerry Lewis expanded his Muscular Dystrophy Association telethon beyond New York City. In the first year, host Phil Markert welcomed the "King of the Cowboys," Roy Rogers, and his wife, Dale Evans. Over the years, local telethon hosts have included Bud Hedinger, George Banks, Nancy Duffy, and Tim Fox. (Courtesy NewsChannel 9.)

Anchorman Solon Gray checks on the construction of the Onondaga County Civic Center, next to the county courthouse in downtown Syracuse. Gray was the centerpiece of a major attempt to upgrade Channel 9 News in the early 1970s. He previously anchored the news at WPIX in New York City. (Courtesy NewsChannel 9.)

Stephen Frazier built a distinguished career as a reporter and anchor at NBC News and CNN. Before that, he was a reporter at 9 News on WNYS-TV. He is seen here on the left, heading to a story in the mid-1970s. (Courtesy NewsChannel 9.)

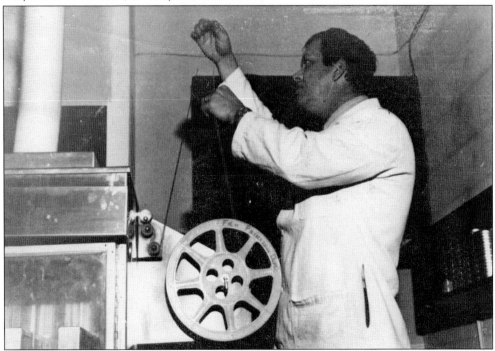

Film editor Dave Stanton looks at the latest images coming off the film processor in Channel 9's news department in Shoppingtown DeWitt. Film was used until the station transitioned to videotape in the mid-1970s. Stanton, a Vietnam veteran, made the transition himself, becoming an electronic news gathering (ENG) cameraman, usually working the night shift. (Courtesy NewsChannel 9.)

Syracuse native Rod Wood was an Army veteran and an established journalist when he was hired from WHEN Radio 62 in April 1976. He immediately brought credibility in the community, and his hiring was the first step in a process that brought 9 News into the ratings war for the first time in the station's history. (Courtesy NewsChannel 9.)

Sportscaster Joe Zone (above) may be as well traveled as any Syracuse broadcaster. In Syracuse, he has worked for Channel 9 (ABC), Channel 3 (NBC), and Channel 68 (Fox). Zone's career has also taken him to the nation's top three markets— New York, Chicago, and Los Angeles—as well as Scranton, Providence, and Hartford. (Courtesy NewsChannel 9.)

Brooklyn native Karin Franklin said she was "looking for the subway" when she came to school at Onondaga Community College. She never found it, but she came to love Central New York and became one of the area's biggest boosters. At Channel 9, she did some consumer reporting and hosted the popular morning show *Open Line,* welcoming guests like Phil Donahue, Sugar Ray Leonard, Libba Cotton, and Eli Wallach. (Courtesy NewsChannel 9.)

Rod Wood is seen here on the set of the 9 *News Hour,* Syracuse's first hour-long newscast, in 1979. The North America weather map served as the backdrop of the multitiered set. Competitors wondered whether Central New York really needed more than 30 minutes of news at one time, and in fact, that first hour-long broadcast lasted just 18 months. WIXT 9 expanded beyond a half-hour format for good in 1983. (Courtesy NewsChannel 9.)

No one chased the news with more determination than Andy Brigham (below). He was known as the most tenacious investigative reporter in Syracuse, and even earned an "Andy Brigham Day" from Syracuse mayor Lee Alexander. WIXT hired Brigham away from WTVH-TV in 1979 to serve as its news director. Brigham was elected to the Syracuse Press Club's Wall of Distinction in 2003. (Courtesy NewsChannel 9.)

Bud Hedinger was one of the most popular personalities in Syracuse, starting as a weatherman at WSYR-TV3 and then hosting *Bowling for Dollars*. WIXT wanted him for weather and a lot more when he crossed town in the late 1970s. He anchored portions of the news, published *Bud Hedinger's Weather Guide*, and covered feature stories around Central New York in *Bud's Journal*. (Courtesy NewsChannel 9.)

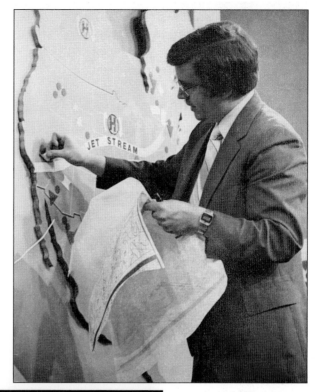

Critic-at-large Doug Brode's movie reviews were among the features added for the *9 News Hour*, Central New York's first hour-long newscast, around 1980. Brode was already a familiar voice on 62 WHEN Radio and in the pages of the *Syracuse New Times*. He also taught film studies at Syracuse University. His Channel 9 ratings system included "Star," "Turkey," and the dreaded "Brode Bomb." (Courtesy NewsChannel 9.)

Charles Anderson and Karen Adams served as cohosts of the public affairs show *Insight* when it launched in 1980. Anderson held a master's degree in television and radio from Syracuse University's Newhouse School. He also served as a Syracuse common councilor from 1985 to 1997. Anderson stayed with *Insight* throughout its 15-year run on weekends on WIXT 9. (Courtesy Charles Anderson.)

Anchor and reporter John Butler stands in front of the *Eyewitness Live* van in this promotional shot from the mid-1980s. Channel 9's first microwave van allowed reporters to file stories live from the field—nearly anywhere in Central New York. Butler was a Central New York native and a Vietnam veteran. He went on to a long career as the news director at CBS-owned KMOX Radio in St. Louis. (Courtesy NewsChannel 9.)

The early evening show *PM Magazine* spent the 1983–1984 season on Channel 9 after a long run on WTVH. Local hosts and feature stories were supplemented by contributions from stations across the country. The *PM* crew on Channel 9 included, from left to right, videographer Vince Spicola, hosts Tim Holton and Diana Lambdin, and executive producer Marirae Dopke, who had hosted the show for a time on WTVH. (Courtesy NewsChannel 9.)

The name *Eyewitness News* returned in the early 1980s, with a twist. Bud Hedinger was installed as the news host, ad-libbing in and out of more traditional reporting from anchors Rod Wood (center) and Sheryl Nathans (left), with Ron Winders (right) on sports. The concept was a ratings flop, and the station quickly returned to a more traditional newscast. (Courtesy NewsChannel 9.)

Rod Wood and Carrie Lazarus have coanchored NewsChannel 9's early evening newscasts for more than 25 years. Wood is a Syracuse native who went into radio after serving in the US Army. Lazarus graduated from Syracuse University's Newhouse School and decided to build her career in Central New York. They were inducted together into the New York State Broadcasters Hall of Fame in 2011. (Courtesy NewsChannel 9.)

Nancy Duffy was one of Central New York's most beloved personalities. She started her career in newspapers before coming to WHEN TV, and later moved to Channel 9. Duffy earned a place on the Syracuse Press Club's Wall of Distinction in 2000, and founded the Syracuse St. Patrick's Parade. She proudly called it "the largest St. Patrick's Parade, per capita, in the world." In this photograph, she's painting the ceremonial green stripe along the parade route. (Courtesy Al Miles.)

The sneakers are on and the ties are off as another Muscular Dystrophy Association telethon comes to an end in Channel 9's Bridge Street studios. A core group of MDA goodwill ambassadors join longtime cohosts Nancy Duffy and Tim Fox in this shot from the late 1980s. From left to right, they are John Noll; Jim, Tracey, and Mike Pappa; Duffy; Jennifer Rittell; Fox; and Howard Klein Jr. (Courtesy MDA/ R.Decker Photos.)

Channel 9's Eyewitness News anchor lineup from the early 1990s had stayed together for more than a dozen years. From left to right, they are meteorologist Dave Eichorn; news anchors Dan Cummings, Rod Wood, and Carrie Lazarus; and sports director Doug Logan. (Courtesy NewsChannel 9.)

Philadelphia-bred investigative reporter Sheryl Nathans joined WIXT in 1983. She was a tenacious investigative reporter who worked her contacts to get the inside story and fought for the underdog. She was also deeply involved in the community. After 13 years at TV9, she moved on to WTVH, and later taught at Syracuse University. She died of cancer in April 2002 at age 55. (Courtesy NewsChannel 9.)

Carrie Lazarus started reporting on health issues at WTVH before she was hired to anchor the early news on WIXT Central New York 9 in 1986. Her nightly *Family Healthcast* health report was one of the first locally produced news features of its kind in the United States. (Courtesy NewsChannel 9.)

Rod Wood and Carrie Lazarus anchor a newscast from the NewsChannel 9 broadcast center outside the Dairy Products Building at the New York State Fairgrounds. Novelist Jay McInerney, the author of *Bright Lights, Big City*, helped build the structure for a little extra cash when he was an instructor in the Syracuse University creative writing program in the early 1980s. (Courtesy NewsChannel 9.)

News expansion pushed forward in the mid-1990s with 9 *News Now*. Maureen Green returned from WTVH to anchor the new 5:00 p.m. newscast with Dan Cummings. Original plans called for Green and Carrie Lazarus to anchor the newscast together, but research found that viewers preferred a male and female anchor team. Green later returned to WTVH for a second stint as its main anchor. (Courtesy NewsChannel 9.)

In the 1990s, Channel 9 took a different approach to morning newscasts, with *Daybreak*. Feature reporter and radio veteran Rick Gary (left) shifted into the anchor seat with veteran reporter Christie Casciano and meteorologist Dave Longley. *Daybreak*'s mix of news and information, community involvement, and humor was a popular choice for Central New York viewers. (Courtesy NewsChannel 9.)

WIXT's state fair site was trashed by 1998's Labor Day storm. Fast-moving thunderstorms moved in early on the morning of September 7. Two people were killed on the fairgrounds, including a vendor in a food stand right near Channel 9. The final day of the fair was canceled, and power was out in some areas of Central New York for more than a week. Damage around the area was estimated at $130 million. (Courtesy NewsChannel 9.)

Sideline reporter Mike Price welcomes nationally known singer-songwriter Amy Grant to the Syracuse St. Patrick's Parade on March 18, 1995. Price's legendary good nature was tested seconds later when Grant's convertible ran over his toe as it drove away. Fortunately, little damage was done. (Courtesy Mike Price.)

It was always good news when Mike Price visited. For more than 25 years, Price closed Channel 9 newscasts with lighthearted features from classrooms and senior centers all over Central New York. He is seen here in April 2003 visiting Bonnie Wenham's fourth-graders at Allen Road Elementary School in North Syracuse. Price was the last original Channel 9 employee remaining when he retired in July 2008. (Courtesy Mike Price.)

Two Syracuse sports icons teamed at WIXT 9 for more than 15 years. Sports director Doug Logan (right) was also the "Voice of the Orange" for 19 years, covering Syracuse University football and basketball on the New York State Radio Network. Jack Morse was best known for honoring hundreds of high school athletes as "Athletes of the Week" at WTVH and WIXT. (Courtesy NewsChannel 9.)

Strong, accurate weather coverage accounts for a good deal of NewsChannel 9's ratings success in the 1990s and 2000s. "Storm Team" meteorologists Jim Teske, Dave Eichorn, and Dave Longley introduced technology such as the Live Doppler 9 and visited classrooms for years with "Weather School" lessons. (Courtesy NewsChannel 9.)

NewsChannel 9 chief meteorologist Dave Eichorn joins Spencer Christian for a national weather update from the New York State Fair on ABC's *Good Morning America* in the late 1990s. It was Christian's second visit to the fair, having broadcast from the fairgrounds previously in 1987. (Courtesy NewsChannel 9.)

Investigative reporter Bill Carey scored a different kind of scoop, interviewing the Drifters during a state fair broadcast in 1997. The hard-nosed Carey rarely covered the entertainment beat, but he talked the group into visiting the NewsChannel 9 broadcast center after an afternoon concert at the nearby Empire Court, and he even talked them into a short performance of their hit, "Up on the Roof." (Courtesy NewsChannel 9.)

Local morning shows had been on hiatus for 20 years or more when NewsChannel 9 launched *Bridge Street* in 2004. Rick Gary and Julie Abbott talked about "everything Central New York," from entertainment to fashion to food to events around the community. Subsequent hosts have included Carrie Lazarus, Dan Cummings, Chris Brandolino, and Kaylea Nixon. (Courtesy NewsChannel 9.)

Four

IN THE PUBLIC INTEREST

Syracuse's commercial television stations had the market all to themselves when educational television joined the pack in late 1965. WCNY Channel 24 faced formidable competition from the big three, but management was determined to make the station a haven for the public's interest with an aggressive local, live, and taped program schedule supplemented by the growing popularity of shows such as *The French Chef* from National Educational Television. Could a meagerly funded black-and-white operation survive when the commercial networks were having their best year ever, in living color no less?

The Onondaga County School Board Association created and initially funded WCNY-TV, which debuted in December 1965. The member-supported station became Central New York's non-commercial station. It was affiliated with National Educational Television (NET), the predecessor of PBS. The Public Broadcasting Council of Central New York, Inc., was created to govern the station.

A booklet pitching the concept spoke of "a handful of people striving to develop within a community an agency which can provide entertainment with enlightenment." In the early days, Channel 24 featured a mélange of local programming in the public interest: classroom instructional programs viewed throughout the county's public schools, how-to shows such as *Straight Down The Middle* offering tips to amateur golfers, and later, fare from the fledgling NET including *The French Chef* with Julia Child and *Mister Rogers' Neighborhood*.

Lacking satellite transmission and coaxial cable links to its affiliates, NET was more a program library service than a full-scale network. Shows were shipped on tape and then bicycled from station to station. Soon, the mix included controversial documentaries on *NET Journal*, thought-provoking drama on *NET Playhouse*, *Firing Line* with William F. Buckley, and imports from Great Britain, including *The Forsyte Saga*. In 1969, *Sesame Street* became a game-changer in children's television. Viewers helped pay the bills by supporting the station's on-air pledge drives, and *TelAuc*, WCNY's most successful fundraiser each year, was part home-shopping network and part on-air carnival. *TelAuc* has become an annual spring tradition for more than 40 years at WCNY.

With a busy local, live schedule for the classroom, WCNY also produced programming that was shared with public stations around New York and around the country. *Ladybug's Garden*, *Pappyland*, live coverage of the Empire State Games and Syracuse University sports, *Fiscal Fitness*, and *The Ivory Tower Half-hour* are just some of the programs produced with support from "viewers like you." It was the pioneering efforts of WCNY-TV's staff to build on a dream and create what has become a cultural institution in Central New York.

Syracuse's public television station was almost called WHTV until a station in nearby Watertown gave up the call letters WCNY. The station's original offices and studio were on Old Liverpool Road in Liverpool. The building was once General Electric's training and industrial film studio. GE's plant in Salina manufactured broadcast equipment and donated monitors and electronic support gear to WCNY. (Courtesy WCNY-TV.)

Syracuse's Robert Weinheimer was one of a team of instructors for a series in 1968 called *Tennis Everyone*. The Westvale students seen here are, from left to right, Michael McInerney, Christine Connors, Rosemary Roike, and Mark Flindt (standing). (Courtesy WCNY-TV.)

Tennis Everyone was designed to teach the fundamentals of tennis play every Saturday. Robert Weinheimer's brother Fred was the second part of the tennis-teaching team and was a well-known Westvale tennis buff. (Courtesy WCNY-TV.)

WCNY's staff announcer Jack Shannon gets ready to interview Nicole Fournier, a hostess for Montreal Expo '67, celebrating "Man and his World." The 1967 studio set shows the fair's logo in the background. Note the a-line polyester knee-length skirt, blouse, blazer, and pillbox hat worn by Fournier, which was inspired by airline stewardess uniforms of the time. (Courtesy WCNY-TV.)

Cinematographer Larry Wood is seen here on assignment at Montreal's Expo '67 with an Arriflex 16-millimeter camera. Wood said he regretted that a tight budget forced him to use black-and-white film to capture the vibrant images of the international exhibition. Back then, as a cinematographer, Wood was responsible for virtually all filming for Channel 24's productions. (Courtesy Larry Wood.)

Black on Black was launched in 1968 with host Charles Anderson. The show was designed to provide a cultural showcase of black artists, craftsmen, journalists, and entertainers. It was sponsored by the Urban League of Onondaga County, which granted Anderson time from his position as education director to produce and host the series. (Courtesy Charles Anderson.)

In 1967, the lineup of shows included the popular children's television series *Ladybug's Garden*, starring Cathy Stampalia as Ladybug and Gay Evans as Sweetie Phew, the mysterious skunk. Instead of a script, the *Garden* cast followed a theme and improvised gestures and words. Director Harry Cunningham said that gave the program vitality and spontaneity. The show was produced for syndication, one year at Channel 9, the other at WCNY. (Courtesy WCNY-TV.)

Cash or Chaos was a late-1960s personal money-management show. Experts from around the country appeared as guests. Here, guest author Morton Shulman (left) talks about investments with Mr. and Mrs. Harold G. McGrath and Cornelius O' Leary, the host of the show. This episode aired on December 15, 1967 at 8:00 p.m. Shulman's book was titled *Anyone Can Make a Million*. (Courtesy WCNY-TV.)

At 9:00 p.m. on November 14, 1967, a one-hour special, *Pills and Pacifiers*, focused on the use and misuse of amphetamines and barbiturates. The studio audience was made up of Liverpool parents, and the show was moderated by Syracuse University's Catherine Covert (far left). Panelists included, from left to right, Dr. Julius Richmond, the dean of Upstate Medical Center; consumer specialist Lois Myer; Syracuse University's Dr. Bernice Wright; and a parent, Mrs. James Provost. (Courtesy WCNY-TV.)

During big political events, WCNY-TV provides viewers with comprehensive coverage. Ron Friedman (right) is seen here in 1968 at the Onondaga County War Memorial behind one of WCNY's General Electric television cameras. That camera's educational mission continued when it was donated to Auburn High School. (Courtesy WCNY-TV.)

A WCNY crew covers an appearance by Republican presidential candidate Richard Nixon and his wife, Pat, at the Onondaga County War Memorial in late October 1968. As WCNY reported, protesters from Syracuse University interrupted Nixon's speech by singing Simon & Garfunkel's "The Sound of Silence." (Courtesy WCNY-TV.)

Channel 24's school services director Elizabeth Nocera guides a student group on a tour of the WCNY-TV facilities in July 1967. Station tours were always very popular with school groups. Nocera has said that children were surprised to see where and how their favorite programs originated. (Courtesy WCNY-TV.)

A major WCNY-TV documentary on the Erie Canal premiered as a fundraiser in 1971 at Loew's State Theater in downtown Syracuse (now the Landmark Theater). *The Ditch That Helped Build America* captured scenes of the Erie Canal and featured vivid descriptions of life in New York State in the 1800s. (Courtesy WCNY-TV.)

Determined to get the best shot for a documentary, Larry Wood risks his safety and his job, putting himself in the middle of a fast-moving river with the station's new, expensive Arriflex BL camera. He was glad his bosses did not find out until after the film was edited. During his career, Wood has won Emmy and Telly Awards and was inducted into the Nevada Broadcasters Hall of Fame. (Courtesy Larry Wood.)

In 1969, WCNY's mobile unit was on location at a neighborhood event, with 16-year-old Mike Clark operating a General Electric PC-12 black-and-white camera mounted atop the mobile unit for a high angle of the activities. Cameras were strategically placed to give viewers the best view of an event. (Courtesy Mike Clark.)

Old Enough to Care was a six-part dramatic series produced by WCNY in 1982 and picked up by PBS for national distribution. Freelance director Hugh Martin is a 10-time Emmy Award nominee. Actors included James McDaniel, who went on to star in *NYPD Blue*. The crew included, from left to right, Charlie Thomas, Dick "Cato" Calagiovanni, Don Deloff, John Duffy, Kevin Rinaldo, Hugh Martin, and Andy Robinson. (Courtesy WCNY-TV.)

Longtime WCNY studio crew chief Steve Parton checks the lighting and sets up a shot for *Old Enough to Care*. The miniseries was shot on location throughout Syracuse, at sites including Arturo's Restaurant, LeMoyne Elementary School, and a storefront on North Salina Street. The dramatic series produced for PBS dealt with age and gender-centered stereotypes. (Courtesy WCNY-TV.)

Carl Eilenberg volunteered many hours for WCNY as a sports broadcaster for basketball, football, hockey, bowling, and even bocce games. Eilenberg would later serve as the mayor of Rome and was the longtime public address voice of the Carrier Dome at Syracuse University. (Courtesy WCNY-TV.)

WCNY's sports coverage was touched by broadcast royalty in the early 1980s when Marty Glickman (left) joined Dave Cohen (right) for WCNY's coverage of the state's high school basketball championship in Rochester. Glickman was one of the nation's first athletes-turned-sportscasters, and he served as the legendary voice of the New York Giants, Knicks, and Jets. Cohen went on to do New York Yankees games on the MSG Network and appear in movies like *Glory Road*. (Courtesy WCNY-TV.)

In the late 1960s and early 1970s, Channel 24 was the only station in town with a remote production truck with multi-camera capability. Hockey fans were delighted with the live coverage of the Syracuse Blazers at the Onondaga County War Memorial. The station's mobile unit carried its own portable microwave transmitter, which beamed a signal to a repeater on the roof of the downtown MONY Building for relay to the station's Liverpool studios. (Courtesy WCNY-TV.)

Richard Calagiovanni (left) sets up a shot with host Rick Young for a series called the *The Genetic Code* in 1976. Calagiovanni joined the team as an intern in 1969 while still a senior at Syracuse University. Affectionately known as "Cato," he became a director in 1971 and a production manager in 1980. He was often compared to *Sesame Street*: painful to replace and impossible to duplicate. (Courtesy WCNY-TV.)

The mighty mobile unit below was designed and constructed by chief engineer Paul Barron. The remote truck steered event coverage in a new direction, with room for three black-and-white GE cameras, a two-inch quad videotape recorder, monitors, a switcher for the director, an audio mixer, engineers, and support equipment. (Courtesy WCNY-TV.)

Ernie Darrh (above) was one of the first engineers hired when the construction of WCNY-TV began in 1965. Here, he is servicing one of the station's second-generation color cameras. The station converted to color in 1971. (Courtesy WCNY-TV.)

Engineer Bob Witkowski is seen here in the tape room in 1985. Copies of locally produced shows are now routinely recorded and archived. In the early days, however, to cut costs, expensive videotape reels were used over and over until they wore out. Like most stations across the country, WCNY saved money but also lost any record of its earliest programs. (Courtesy WCNY-TV.)

The stars of one of the station's first televised auction shows, *TelAuc*, perform what they called their "Busby Berkeley" number. *TelAuc* continues to be WCNY-TV's single largest fundraiser of the year, thanks to auction items donated by businesses and individuals across Central New York. The televised auction celebrated its 45th year in 2013. (Courtesy WCNY-TV.)

The first *TelAuc* used Marconi black-and-white studio cameras purchased from CBS. The rest of the auctions were in color. Those early monochrome cameras came with some fascinating history. They came from New York City, where they captured the historic appearances of the Beatles on *The Ed Sullivan Show* in 1964. (Courtesy WCNY-TV.)

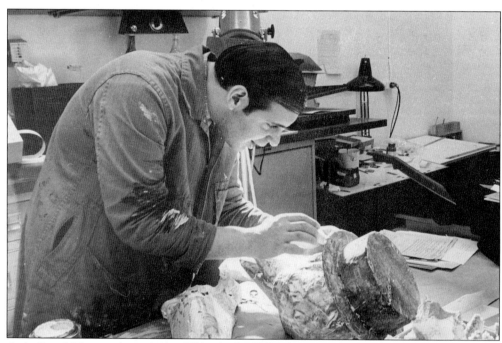

Artist Steve Meltzer works on the first *TelAuc* bird in 1968. The mascot for the telethon made appearances around town to drum up excitement for the annual fundraiser. (Courtesy WCNY-TV.)

"He's everywhere! He's everywhere!" Here, the *TelAuc* bird is a big hit with kids at the ballpark. This stop at MacArthur Stadium on the city's north side was one of many public appearances before the big auction. (Courtesy WCNY-TV.)

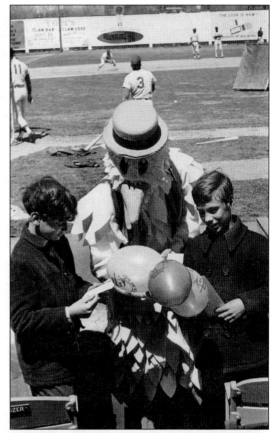

The auction to raise money for WCNY-TV relied on volunteers. Corporations like Crucible Steel sent dozens of employees (below) to answer the phones and take pledges from viewers across Central New York. (Courtesy WCNY-TV.)

Popular radio personalities Bill Merchant (left) and Ted Downes brought their humor and cast of characters to the *TelAuc* stage to host and help auction off special items in the early 1980s. *TelAuc* would often invite local celebrities to appear as guest auctioneers. (Courtesy WCNY-TV.)

Nancy Roberts was part of the talent lineup for *TelAuc* in the early 1980s. The marathon fundraisers required station talent to take shifts and be upbeat. Many opted to wear sneakers during their long shifts on the hard concrete studio floor. (Courtesy WCNY-TV.)

Every August, the *Bluegrass Ramble Picnic* unfolds in the country. This is a photograph of the big party in 1975 at a farm in Cato. The event has become a favorite tradition and draws a huge crowd every year. (Courtesy WCNY-TV.)

Longtime *Bluegrass Ramble* host Bill Knowlton is a tireless promoter of bluegrass. In 2011, he received the International Bluegrass Music Association's Distinguished Achievement Award in Nashville. (Courtesy WCNY-TV.)

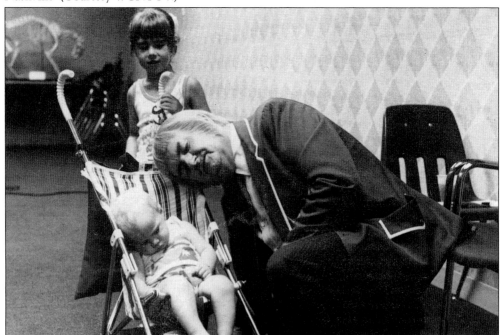

Captain Kangaroo had moved from CBS to PBS in August 1987, bringing the Captain himself to Syracuse. Bob Keeshan made several appearances on behalf of WCNY-TV, including this one at the Rosamond Gifford Zoo. (Courtesy WCNY-TV.)

WCNY's first *Studio Club* event drew talk-show heavyweight Dick Cavett on February 9, 1981, at the Hotel Syracuse's Persian Terrace. Cavett was hosting a PBS show at the time, and appeared with Syracuse Symphony Orchestra maestro Christopher Keene (center) and Syracuse Stage founder Arthur Storch (right). (Courtesy WCNY-TV.)

WCNY-TV often took its talk shows on the road. Syracuse University basketball coach Jim Boeheim was the first guest on *Profile*, hosted by Caroline Coley, in 1990. The interview was conducted at the Carrier Dome on the Syracuse University campus. (Courtesy WCNY-TV.)

Few locally produced shows make it past 10 years, but *Financial Fitness*, WCNY's longest-running series, celebrated its 21st season in 2012. The original production team is seen in this 1992 staff shot, with director Linda Marie Randulfe, cohost Tory McCleod, host Dan Pluff, and executive producer Jack Neal. Pluff, a financial advisor, created the show and served as host until 2012. (Courtesy WCNY-TV.)

In 1988, a new, more comfortable set was built for *Good Afternoon*, with hosts Ron Schuver and Rochelle Cassella. The talk show aired live, and took phone calls from viewers about everything, including politics, sports, and whatever else viewers had on their minds. The show even included "open-air" segments to allow viewers to vent. (Courtesy WCNY-TV.)

WCNY-TV supported the local arts, and in turn, local artists supported the station. Central New York artist Hall Groat, the brother of NewsChannel 9 anchor Rod Wood, is seen here. Groat helped the station raise thousands of dollars by donating his paintings for the annual art auction in the 1980s. (Courtesy WCNY-TV.)

Gerard Moses starred as the man with the answers in *Behind the Wheel*, WCNY-TV's 1979–1980 contract production for the State of New York School Bus Driver Training Program. (Courtesy WCNY-TV.)

The star of *Pappyland*, cartoonist Michael Cariglio, leads the WCNY-TV contingent at the 1995 Syracuse St. Patrick's Parade in downtown Syracuse. Cariglio starred as Pappy Drewitt, who encouraged kids to be good and enjoy the arts. The live-action children's show was produced in part at WCNY and aired nationally from 1996 to 1999 on the TLC cable network. (Courtesy WCNY-TV.)

Five

NEW CHOICES

More local choices became available in 1987, when WSYT Channel 68 signed on the air. Two other UHF stations later joined the Fox affiliate: WSTQ, now branded CW6, and what is now WNYS My 43.

The local television landscape changed once again in November 2003 with the launch of Central New York's first and only 24-hour news channel, owned and operated by Time Warner Cable. News 10 Now was three years in the making, from the initial decision to build the channel to its actual launch. Two of those years involved the renovation and restoration of the channel's headquarters, the historic New York Central train station on Erie Boulevard East. The dilapidated structure was better known as the former Greyhound bus station, and had been vacant for several years before the decision was made to purchase the property and invest in the multimillion-dollar reconstruction project.

News 10 Now initially served 345,000 cable subscribers in the greater Syracuse area and the Watertown/North Country region. Over time, another 230,000 customers were added across a broader chunk of central and upstate New York and northern Pennsylvania. A completely separate second program feed, customized for Southern Tier regional viewers, was launched in February 2007.

News 10 Now rebranded as YNN (Your News Now) in early 2010. The 70-person YNN team is based in Syracuse, has seven bureaus from Potsdam to Corning, and gathers news with a heavy emphasis on regional reporting. The channel pioneered the video-journalist concept locally, equipping its 18 reporters with cameras and gear to "one-man band" their stories. The Syracuse-based YNN channels share an anchor team, creative services, news production, and a master control facility with its sister channels in Albany and the Hudson Valley. Those functions are all based in Albany.

The Syracuse station, in turn, is the weather and website hub for all six of the company's upstate news channels, with eight "Weather on the 10s" meteorologists and four web producers based on Erie Boulevard.

Time Warner Cable first announced plans to create the Syracuse news channel in 2000, but progress was slow until the cable company purchased the train station property on Erie Boulevard East and began the remediation of the property in 2001. (Courtesy Ron Lombard.)

General manager Ron Lombard joined Time Warner Cable in February 2002 with the goal of launching the channel late that same year. But there were delays due to the enormity of the building renovation and restoration, and it was another 20 months before the project was completed. (Courtesy Ron Lombard.)

The news channel finally took to the cable system with its original name News 10 Now at 2:00 p.m. on Friday, November 7, 2003. Syracuse mayor Matt Driscoll and Onondaga County executive Nick Pirro joined Time Warner Cable Syracuse Division president Mary Cotter in the control room as general manager Ron Lombard flipped the switch to take the channel live to the cable system. (Courtesy Klineberg Photography.)

News 10 Now general manager Ron Lombard hands a commemorative block to Time Warner Cable's Mary Cotter. The blocks were engraved with the news channel's logo and launch date and were made from pieces of the building's original limestone facing. They were presented to several people who were instrumental in the channel's successful launch. (Courtesy Klineberg Photography.)

The station's newsroom in Syracuse is seen here. News 10 Now was rebranded as YNN (Your News Now) in early 2010. (Courtesy Klineberg Photography.)

Employees and guests at News 10 Now's 2003 launch party get a tour of the channel's state-of-the-art, fully digital production control room. (Courtesy Klineberg Photography.)

YNN's "Weather on the 10s" studio is the home for eight Syracuse-based meteorologists, who provide weathercasts for all seven of Time Warner Cable's New York State 24-hour news channels. (Courtesy Klineberg Photography.)

Three-time Emmy winner Bill Carey takes a stab at cable news. Carey was a radio news director at WHEN and WSYR Radio, the news director for WTVH-TV, and a reporter for WIXT-TV before joining News 10 Now's original team as its senior reporter. The YNN staff helped Bill celebrate his illustrious career on the 40th anniversary of his first day as a Central New York journalist in April 2011. (Courtesy Ron Lombard.)

YNN morning news anchors JoDee Kenney (left) and Julie Chapman were the original solo morning anchors of the Time Warner Cable news channels in Syracuse and Albany, respectively. When anchoring and technical news production was consolidated into the Albany facility in 2005, Kenny and Chapman became the anchor team for both channels. (Courtesy Klineberg Photography.)

News 10 Now's original sports team included, from left to right, Will Medina, sports director Mark Larson, and Chris Watson. YNN has since grown to a six-person sports department, which covers an area of 15,000 square miles. (Courtesy Klineberg Photography.)

News 10 Now's original "Weather on the 10s" team of meteorologists included, from left to right, John DiPasquale, Carrie Cheevers, Michael Gouldrick, and chief meteorologist Jack Church. (Courtesy Klineberg Photography.)

Betsy Sykes and Chuck Plumpton were the anchors for *Fox 68 News at 10* on WSYT Channel 68 when it debuted in 1995. The broadcast was Syracuse's first 10:00 p.m. newscast. Over the years, production of the newscast was contracted to WSTM and later WTVH. (Courtesy of *Syracuse New Times*.)

Anchor Matt Mulcahy is seen here on the set of the 10:00 p.m. news carried on WSTQ-TV. The station and WSTM-TV were then owned by Raycom Media and had launched a 10:00 p.m. news program the year before to compete with the Fox station's 10:00 p.m. news. (Courtesy of *Syracuse New Times*.)

Magazine 13 originated on Newchannel's suburban cable system (now Time Warner Cable) and aired in Syracuse and its suburbs. The show featured guests and features about the people of Central New York. From left are the show's hosts, Ron Curtis Jr., Nancy Roberts, and Jack Morse. The program ran from 1989 to 2003. (Courtesy of *Syracuse New Times*.)

Rough Times Live, airing on Time Warner Cable 13, was produced by *The Media Unit,* a Syracuse-based performance group that does programs by, about, and for teens. The program is still seen weekly on the cable system's public-access channel. In this 1991 photograph, Walt Shepperd, the director of *The Media Unit,* guides his young performers in a production. (Courtesy of *Syracuse New Times.*)

Discover Thousands of Local History Books Featuring Millions of Vintage Images

Arcadia Publishing, the leading local history publisher in the United States, is committed to making history accessible and meaningful through publishing books that celebrate and preserve the heritage of America's people and places.

Find more books like this at
www.arcadiapublishing.com

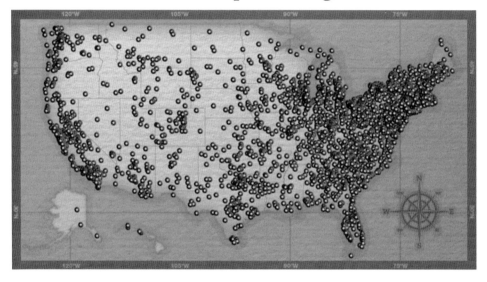

Search for your hometown history, your old stomping grounds, and even your favorite sports team.